CREEPY CRAWL

SCORPIONS

by Nessa Black

AMICUS | AMICUS INK

stinger

legs

Look for these words and pictures as you read.

pincer

babies

What is on the rock?

It is a scorpion.

It is ready to hunt.

Do not get too close.

It can pinch.

It can sting.

Look at its stinger.
It is sharp.
It has venom.

stinger

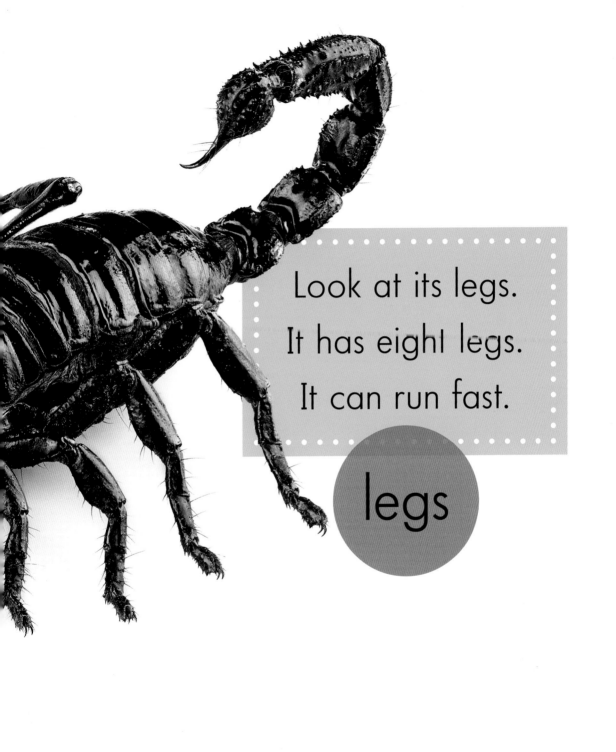

Look at its legs.
It has eight legs.
It can run fast.

legs

Look at its pincer.
A scorpion has two.
They grab food.

pincer

Look at its babies.
They have soft shells.
They ride on their mom
for 1 to 3 weeks.

babies

There are many kinds
of scorpions.
How many do you see?

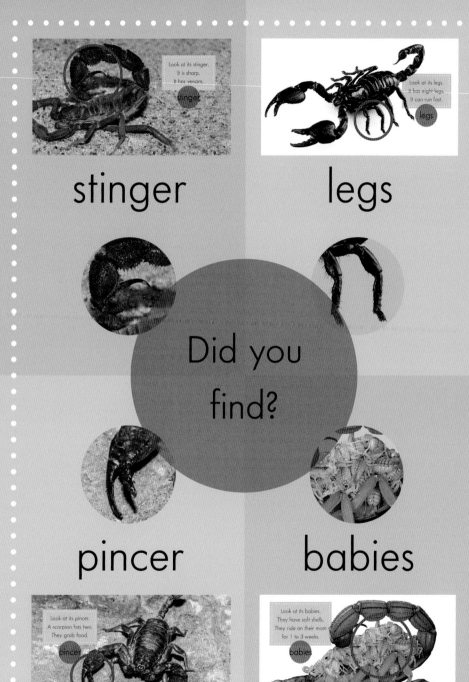

stinger

legs

Did you find?

pincer

babies

Spot is published by Amicus and Amicus Ink
P.O. Box 1329, Mankato, MN 56002
www.amicuspublishing.us

Library of Congress Cataloging-in-Publication Data
Names: Black, Nessa, author.
Title: Scorpions / by Nessa Black.
Description: Mankato, MN : Amicus/Amicus Ink, [2019] |
  Series: Spot. creepy crawlies | Audience: K to grade 3.
  | Description based on print version record and CIP data
  provided by publisher; resource not viewed.
Identifiers: LCCN 2017046895 (print) | LCCN 2017050587
  (ebook) | ISBN 9781681515779 (pdf) | ISBN
  9781681515397 (library binding) | ISBN 9781681523774
  (pbk.)
Subjects: LCSH: Scorpions–Juvenile literature.
Classification: LCC QL458.7 (ebook) | LCC QL458.7 .B58
  2019 (print) | DDC 595.4/6–dc23
LC record available at https://lccn.loc.gov/2017046895

Printed in China

HC 10 9 8 7 6 5 4 3 2 1
PB 10 9 8 7 6 5 4 3 2 1

Wendy Dieker and Alissa Thielges,
  editors
Deb Miner, series designer
Kazuko Collins, book designer
Holly Young, photo researcher

Photos by Shutterstock/Aleksey
Stemmer cover, 14; Alamy/R Kawka
4–5, Domiciano Pablo Romero Franco
8–9; iStock/praisaeng 1; Minden/
Piotr Naskrecki 3, Ingo Arndt 12–13;
Shutterstock/arka38 ¥4, Fabrizo Conte
10–11, Piyathep 14, Sunshine Studio 14,
Vydunas 6–7

SCORPIONS